Samuel Henshaw

The Entomological Writings of Dr. Alpheus Spring Packard

Samuel Henshaw

The Entomological Writings of Dr. Alpheus Spring Packard

ISBN/EAN: 9783337088798

Printed in Europe, USA, Canada, Australia, Japan

Cover: Foto ©berggeist007 / pixelio.de

More available books at **www.hansebooks.com**

U. S. DEPARTMENT OF AGRICULTURE.

DIVISION OF ENTOMOLOGY.

BULLETIN No. 16.

THE

ENTOMOLOGICAL WRITINGS

OF

DR. ALPHEUS SPRING PACKARD.

BY

SAMUEL HENSHAW.

WASHINGTON:
GOVERNMENT PRINTING OFFICE.
1887.

LETTER OF SUBMITTAL.

U. S. Department of Agriculture,
Division of Entomology,
Washington, D. C., July 5, 1887.

Sir: I have the honor to submit for publication Bulletin No. 16 of this Division, being a list of the entomological writings of Dr. A. S. Packard, with systematic and general index, prepared by Mr. Samuel Henshaw. Dr. Packard has been so long and favorably known as a writer upon insects both in their structural, biologic, and economic relations, and has been for so many years connected with Government entomological work, that this Bulletin will be welcomed by all interested in the subject and of great aid in the divisional work.

Respectfully,

C. V. RILEY,
Entomologist.

Hon. Norman J. Colman,
Commissioner of Agriculture.

3

THE ENTOMOLOGICAL WRITINGS OF ALPHEUS SPRING PACKARD.

By SAMUEL HENSHAW.

Alpheus Spring Packard was born in Brunswick, Me., February 19, 1839. His father was Alpheus Spring Packard, D. D., for over sixty years a professor in Bowdoin College. His mother was Frances E. Appleton, daughter of Rev. Jesse Appleton, president of Bowdoin College. After graduating from Bowdoin College in 1861, he spent three years at the Cambridge Museum of Comparative Zoology as a student of Prof. L. Agassiz. For a part of one year (1863–'64) he was the private assistant of Professor Agassiz.

Two summers (those of 1860 and 1864) were passed upon the coast of Labrador, where collections of marine invertebrates, insects, and quaternary fossils were accumulated for future investigations. In 1861–'62 he was assistant to the Maine Geological Survey. In 1864 he took the degree of Doctor of Medicine at the Maine Medical School. In September of the same year Dr. Packard was commissioned assistant surgeon First Maine Veteran Volunteer Infantry, and served in the Sixth Corps until mustered out with the regiment in July, 1865. In 1865–'66 he was acting custodian and librarian of the Boston Society of Natural History. Dr. Packard spent eleven years (1867–'78) in Salem. Appointed in 1867 one of the curators of the Peabody Academy, he was for about two years (1877–'78) the director of its museum. At Salem he established a summer school of biology, and in March, 1868, the first number of the American Naturalist was issued. Dr. Packard was one of the originators of this magazine, and for twenty years its editor-in-chief.

In 1867 he married Elizabeth Derby, daughter of Samuel B. Walcott of Salem, and has had four children, of whom a son and two daughters are living. As lecturer or instructor Dr. Packard has been connected with the Anderson School of Natural History, Bowdoin College, and the Maine and Massachusetts State Agricultural Colleges; as assistant he has been attached to the Kentucky Geological Survey, to Hayden's United States Geological Survey of the Territories, and to the United States Fish Commission. When in search of material for his studies

Dr. Packard has visited many parts of the United States and Mexico, and has dredged upon the coast of Labrador, in the Gulf of Maine, in Massachusetts and Buzzards Bays, off Beaufort, N. C., and upon the coast of Florida.

In 1871–'73 he served as State entomologist of Massachusetts, and from 1877–'82 was a member of the United States Entomological Commission. In 1878 he accepted the professorship of Zoology and Geology in Brown University, and still retains the position.

Dr. Packard was elected a member of the National Academy in 1872 and honorary member of the Entomological Society of London in 1884, and at home and abroad a number of societies have elected him to membership.

The entomological writings of Dr. A. S. Packard, recorded in Part I of the present list, form but a portion of his contributions to science. His memoirs in other branches in range cover the field of zoology, with occasional papers in allied sciences, and bear testimony alike to his versatility and the energy of his mind.

As a rule only the original place of publication is recorded, though a few reprints and reviews are included, as of possible value in case the original is inaccessible.

Dr. Packard's contributions to the natural history of *Limulus* are included in the present bibliography, because much of the discussion concerning the anatomy, genealogy, &c., of this animal bears directly upon the *Arachnida* and other *Arthropoda*.

Part II contains a systematic list of the new names proposed by Dr. Packard, and I have endeavored to note the collection containing the type, but in many cases have been unable to do so.

A number of the types noted as present in the collection of the Museum of Comparative Zoology are, however, in a very poor state of preservation, so that a word of explanation should be added.

The bulk of Dr. Packard's types were accumulated during his connection with, and formed part of the collection of, the Peabody Academy of Science at Salem.

From the year 1880 the Academy was without the services of an entomological assistant, so that the collections, " in spite of what care could be given them, were rapidly going to ruin," when, fortunately, in 1885, their valuable collections of insects were deposited without conditions in the museum at Cambridge, where their preservation is assured.

Dr. Packard has aided me throughout the preparation of the list, and I am indebted to Messrs. Edwards, Fernald, Hulst, Riley, and Smith for assistance in determining the value and position of many of the *Lepidoptera*. Mr. Howard has rendered a similar service with some of the parasitic *Hymenoptera*.

PART I.

CHRONOLOGICAL CATALOGUE.

1.

1861. PACKARD, ALPHEUS S. Entomological report on the Army-worm and Grain Aphis. <6th Ann. Rept. Me. Bd. Agric., 1861, pp. 130-145.

2.

1861. PACKARD, ALPHEUS S. Report on the Insects collected on the Penobscot and Alleguash Rivers during August and September, 1861. <6th Ann. Rept. Me. Bd. Agric., 1861, pp. 373-376.

3.

1862. PACKARD, ALPHEUS S. How to observe and collect Insects. <2d Ann. Rept. Nat. Hist. and Geol. Me., 1862, pp. 143-219, figs. Separate: Augusta, 1863, pp. 79, figs.

4.

1863. PACKARD, ALPHEUS S. On synthetic types in Insects. <Bost. Journ. Nat. Hist., 1863, v. 7, pp. 590-603, figs.

5.

1864. PACKARD, ALPHEUS S. [Note on Stylops childreni.] <Proc. Ent. Soc. Phil., 1864, v. 3, pp. 44-45.

6.

1864. PACKARD, ALPHEUS S. Synopsis of the Bombycidæ of the United States <Proc. Ent. Soc. Phil., 1864, v. 3, pp. 97-130; 331-396.

7.

1864. PACKARD, ALPHEUS S. Note on the family Zygœnidæ. <Proc. Essex Inst., 1864, v. 4, pp. 7-47, pl. 1-2.

8.

1864. PACKARD, ALPHEUS S. The Humble-bees of New England and their parasites; with notices of a new species of Anthophorabia and a new genus of Proctotrupidæ. <Proc. Essex Inst., 1864, v. 4, pp. 107-140, pl. 3.

7

9.

1864. PACKARD, ALPHEUS S. Report on the collection of Iusects for 1863. < *Rept. Mus. Comp. Zool.*, 1864, pp. 36–44.

10.

1865. PACKARD, ALPHEUS S. Notes on two Ichneumons parasitic on *Samia columbia*. <*Proc. Bost. Soc. Nat. Hist.* 1865, v. 9, pp. 345–346.

11.

1866. PACKARD, ALPHEUS S. Observations on the development and position of the *Hymenoptera* with notes on the morphology of Insects. <*Proc. Bost. Soc. Nat. Hist.*, 1866, v. 10, pp. 279–295, figs; *Ann. and Mag. Nat. Hist.*, 1866, ser. 3, v. 18, pp. 82–99, figs.

12.

1866. PACKARD, ALPHEUS S. Revision of the fossorial *Hymenoptera* of North America. <*Proc. Ent. Soc. Phil.*, 1866, v. 6, pp. 39–115 ; 1867, v. 6, pp. 353–444.

13.

1866. PACKARD, ALPHEUS S. On certain entomological speculations—a review. <*Proc. Ent. Soc. Phil.*, 1866, v. 6, pp. 209–218.

14.

1867. PACKARD, ALPHEUS S. Insects and their allies. <*Amer. Nat.*, 1867, v. 1, pp. 73–84, figs.

15.

1867. PACKARD, ALPHEUS S. Wasps as marriage priests of plants. <*Amer. Nat.* 1867, v. 1, pp. 105–106, fig.

16.

1867. PACKARD, ALPHEUS S. The Insects of early spring. <*Amer. Nat.*, 1867, v. 1, pp. 110–111.

17.

1867. PACKARD, ALPHEUS S. The Insects of May. <*Amer. Nat.*, 1867, v. 1, pp. 162–164, figs.

18.

1867. PACKARD, ALPHEUS S. The Insects of June. <*Amer. Nat.*, 1867, v. 1, pp. 220–224, figs.

19.

1867. PACKARD, ALPHEUS S. The Red-legged Grasshopper. <*Amer. Nat.*, 1867, v. 1, pp. 271–272.

20.

1867. PACKARD, ALPHEUS S. The Insects of July. <*Amer. Nat.*, 1867, v. 1, pp. 277–279, figs.

21.

1867. PACKARD, ALPHEUS S. The Dragon-fly. <*Amer. Nat.*, 1867, v. 1, pp. 304–313, pl. 9, figs ; *Science Gossip*. 1867, pp. 225–227.

22.

1867. PACKARD, ALPHEUS S. The Insects of August. <*Amer. Nat.*, 1867, v. 1, pp. 327-330, figs.

23.

1867. PACKARD, ALPHEUS S. The home of the Bees. <*Amer. Nat.*, 1867, v. 1, pp. 364-378; 1868, v. 1, pp. 596-606, pl. 10, figs.

24.

1867. PACKARD, ALPHEUS S. The eggs of the Dragon-fly. <*Amer. Nat.*, 1867, v. 1, p 391.

25.

1867. PACKARD, ALPHEUS S. Insects in September. <*Amer. Nat.*, 1867, v. 1, pp. 391-392.

26.

1867. PACKARD, ALPHEUS S. The Clothes-moth. <*Amer. Nat.*, 1867, v. 1, pp. 423-427, figs.

27.

1867. PACKARD, ALPHEUS S. [Review of] Lubbock: Development of *Chlœon.* <*Amer. Nat.*, 1867, v. 1, pp. 428-431.

28.

1867. PACKARD, ALPHEUS S. The horned *Corydalus.* <*Amer. Nat.*, 1867, v. 1, pp. 436-437, figs.

29.

1867. PACKARD, ALPHEUS S. The Tiger-beetle. <*Amer. Nat.*, 1867, v. 1, pp. 552-554, figs.

30.

1867. PACKARD, ALPHEUS S. View of the lepidopterous fauna of Labrador. <*Proc. Bost. Soc. Nat. Hist.*, 1867, v. 11, pp. 32-63.

31.

1867. PACKARD, ALPHEUS S. [Increasing distribution of the Canker-worm.] <*Proc. Bost. Soc. Nat. Hist.*, 1867, v. 11, p. 88.

32.

1867. PACKARD, ALPHEUS S. Materials for a monograph of the *Phalænidæ* of North America. <*Proc. Bost. Soc. Nat. Hist.*, 1867, v. 11, pp. 102-103.

33.

1867. PACKARD, ALPHEUS S. [On the larva of *Scenopinus*?] <*Proc. Essex Inst.*, 1867, v. 5, p. 94, fig.

34.

1867. PACKARD, ALPHEUS S. [Larva of salt-water *Chironomus.*] <*Proc. Essex Inst.*, 1867, v. 5, p. 187.

35.

1868. PACKARD, ALPHEUS S. The Insect fauna of the summit of Mount Washington as compared with that of Labrador. <*Amer. Nat.*, 1868, v. 1, pp. 674-676.

36.

1868. PACKARD, ALPHEUS S. On the development of a Dragon-fly, *Diplax*. <*Amer. Nat.*, 1868, v. 1, pp. 676-680, figs.

37.

1868. PACKARD, ALPHEUS S. Are Bees injurious to fruit? <*Amer. Nat.*, 1868, v. 2, p. 52.

38.

1868. PACKARD, ALPHEUS S. Apiphobia. <*Amer. Nat.*, 1868, v. 2, pp. 108-109.

39.

1868. PACKARD, ALPHEUS S. Entomological calendar [for April.] <*Amer. Nat.*, 1868, v. 2, pp. 110-111, figs.

40.

1868. PACKARD, ALPHEUS S. Fossil Insects. <*Amer. Nat.*, 1868, v. 2, p. 163, fig.

41.

1868. PACKARD, ALPHEUS S. Entomological calendar [for May.] <*Amer. Nat.*, 1868, v. 2, pp. 163-165, figs.

42.

1868. PACKARD, ALPHEUS S. The parasites of the Honey-bee. <*Amer. Nat.*, 1868, v. 2, pp. 195-205, pl. 4-5.

43.

1868. PACKARD, ALPHEUS S. Entomological calendar [for June.] <*Amer. Nat.*, 1868, v. 2, pp. 219-221, figs.

44.

1868. PACKARD, ALPHEUS S. Insects living in the sea. <*Amer. Nat.*, 1868, v. 2, pp. 277-278, figs.

45.

1868. PACKARD, ALPHEUS S. Salt-water Insects. <*Amer. Nat.*, 1868, v. 2, pp. 329-330.

46.

1868. PACKARD, ALPHEUS S. Entomological calendar [for August.] <*Amer. Nat.*, 1868, v. 2, pp. 331-334, figs.

47.

1868. PACKARD, ALPHEUS S. [Note on Fire-flies.] <*Amer. Nat.*, 1868, v. 2, pp. 432-433, figs.

48.

1868. PACKARD, ALPHEUS S. [Note on the Moose-tick.] <*Amer. Nat.*, 1868, v. 2, p. 559, figs.

49.

1868. PACKARD, ALPHEUS S. The embryology of *Libellula* (*Diplax*), with notes on the morphology of Insects, and the classification of the *Neuroptera*. <*Proc. Amer. Assoc. Adv. Sci.*, 1868, v. 16, pp. 153-154.

50.

1868. PACKARD, ALPHEUS S. The Insect fauna of the summit of Mount Washington as compared with that of Labrador. <*Proc. Amer. Assoc. Sci.*, 1868, v. 16, pp. 154-158.

51.

1868. PACKARD, ALPHEUS S. On the development of a Dragon-fly (*Diplax*.) <*Proc. Bost. Soc. Nat. Hist.*, 1868, v. 11, pp. 365-372, figs.

52.

1868. PACKARD, ALPHEUS S. [Note on salt-water Insects.] <*Proc. Bost. Soc. Nat. Hist.*, 1868, v. 11, pp. 387-388.

53.

1868. PACKARD, ALPHEUS S. On the structure of the ovipositor and homologous parts in the male insect. < *Proc. Bost. Soc. Nat. Hist.*, 1868, v. 11, pp. 393-399, figs.

54.

1869. PACKARD, ALPHEUS S. [Review of] Meek, Worthen, and Scudder: Articulate fossils from the coal. <*Amer. Nat.*, 1869, v. 3, pp. 45-46, fig.

55.

1869. PACKARD, ALPHEUS S. [Review of] Claparede: Studien an Acariden. <*Amer. Nat.*, 1869, v. 3, pp. 490-493, pl. 8.

56.

1869. PACKARD, ALPHEUS S. A chapter on Flies. <*Amer Nat.*, 1869, v. 2, pp. 586-596; 638-644, pl. 12-13, figs.

57.

1869. PACKARD, ALPHEUS S. Case-worms. <*Amer. Nat.*, 1869, v. 2, pp. 160-161, figs.

58.

1869. PACKARD, ALPHEUS S. A chapter on Mites. <*Amer. Nat.*, 1869, v. 3, pp. 364-373; 448, pl. 6, figs.

59.

1869. PACKARD, ALPHEUS S. The Salt Lake *Ephydra*. <*Amer. Nat.*, 1869, v. 3, p. 391.

60.

1869. PACKARD, ALPHEUS S. On Insects inhabiting salt water. <*Proc. Essex Inst.*, 1869, v. 6, pp. 41-51, figs.

61.

1869. PACKARD, ALPHEUS S. The characters of the lepidopterous family *Noctuidæ*. < *Proc. Port. Soc. Nat. Hist.*, 1869, v. 1, pp. 153-156.

62.

1869. PACKARD, ALPHEUS S. Report of the Curator of *Articulata*, [Peab. Acad. Sci.] < *1st Ann. Rept. Trustees Peab. Acad. Sci.*, 1869, pp. 52–56.

63.

1869. PACKARD, ALPHEUS S. List of hymenopterous and lepidopterous Insects collected by the Smithsonian expedition to South America, under Prof. James Orton. <*1st Ann. Rept. Trustees Peab. Acad. Sci.*, 1869, pp. 56–69.

64.

1869. PACKARD, ALPHEUS S. Guide to the study of Insects, and a treatise on those injurious and beneficial to crops. <*Salem*, 1869, pp. 8 + 702, pl. 1-11, figs.
  a. 2d edition, Salem, 1870.
  b. 3d edition, Salem, 1872.
  c. 4th edition, Salem, 1874.
  d. 5th edition, New York, 1876.
  e. 6th edition, New York, 1878.
  f. 7th edition, New York, 1880.
  g. 8th edition, New York, 1884, pp. 8 + 715, pl. 1-15, figs.

65.

1869. PACKARD, ALPHEUS S. Record of American Entomology for the year 1868. <*Salem*, 1869, pp. 6 + 52.

66.

1870. PACKARD, ALPHEUS S. [Note on *Epeira riparia* and *E. cancer*.] <*Amer. Nat.*, 1870, v. 3, p. 616.

67.

1870. PACKARD, ALPHEUS S. Certain parasitic Insects. <*Amer. Nat.*, 1870, v. 4, pp. 83–99, pl. 1, figs.

68.

1870. PACKARD, ALPHEUS S. A few words about Moths. <*Amer. Nat.*, 1870, v. 4, pp. 225–229, pl. 2.

69.

1870. PACKARD, ALPHEUS S. Embryology of *Limulus polyphemus*. <*Amer. Nat.*, 1870, v. 4, pp. 498–502, figs. *Quart. Journ. Micros. Sci.*, 1871, ser. 2, v. 11, pp. 263–267.

70.

1870. PACKARD, ALPHEUS S. [*Pieris rapæ* in New Jersey.] <*Amer. Nat.*, 1870, v. 4 p. 576.

71.

1870. PACKARD, ALPHEUS S. The borers of certain shade-trees. <*Amer. Nat.*, 1870, v. 4, pp. 588–594, figs.

72.

1870. PACKARD, ALPHEUS S. [Review of] Riley: Second Missouri Report. <*Amer. Nat.*, 1870, v. 4, pp. 610–615, figs.

73.

1870. PACKARD, ALPHEUS S. The caudal styles of Insects sense organs, *i. e.*, abdominal antennæ. <*Amer. Nat.*, 1870, v. 4, pp. 620–621.

74.

1870. PACKARD, ALPHEUS S. A remarkable Myriapod: [*Pauropus Lubbockii.*] <*Amer. Nat.*, 1870, v. 4, p. 621.

75.

1870. PACKARD, ALPHEUS S. Abdominal sense organs in a Fly. <*Amer. Nat.*, 1870, v. 4, pp. 690–691.

76.

1870. PACKARD, ALPHEUS S. [The Currant Saw-fly.] <*Bull. Essex Inst.*, 1870, v. 2, pp. 93–95, figs.

77.

1870. PACKARD, ALPHEUS S. List of *Coleoptera* collected by A. S. Packard, jun., at Caribou Island, Labrador, Straits of Belle Isle. <*Can. Ent.*, 1870, v. 2, p. 119.

78.

1870. PACKARD, ALPHEUS S. New or little known injurious Insects. <17*th Ann. Rept. Sec. Mass. Bd. Agric.*, 1870, pp. 235–263, pl. 1, figs. Separate : 1870, pp. 31, pl. 1, figs. See *Amer. Nat.*, 1871, v. 4, pp. 684–687, pl. 6, figs.)

79.

1870. PACKARD, ALPHEUS S. Record of American Entomology for the year 1869. <*Salem*, 1870, pp. 5 + 62.

80.

1871. PACKARD. ALPHEUS S. On the Insects inhabiting salt water, No. 2. <*Amer. Journ. Sci.*, 1871, ser 3, v. 1, pp. 100–110, figs. · *Ann. and Mag. Nat. Hist.*, 1871, ser. 4, v. 7, pp. 230–240.

81.

1871. PACKARD, ALPHEUS S. Morphology and ancestry of the King Crab. <*Amer. Nat.*, 1871, v. 4, pp. 754–756.

82.

1871. PACKARD, ALPHEUS S. The ancestry of Insects. <*Amer. Nat.*, 1871, v. 4, p. 756.

83.

1871. PACKARD, ALPHEUS S. [Review of] Ganin: The early stages of Ichneumon parasites. <*Amer. Nat.*, 1871, v. 5, pp. 42–52, figs.

84.

1871. PACKARD, ALPHEUS S. Bristle-tails and Spring-tails. <*Amer. Nat.*, 1871, v. 5, pp. 91–107, pl. 1, figs.

85.

1871. PACKARD, ALPHEUS S. [Review of] Adair: Annals of Bee culture for 1870. <*Amer. Nat.*, 1871, v. 5, pp. 113–115.

86.

1871. PACKARD, ALPHEUS S. [Review of] Eaton: Monograph on the *Ephemeridæ*. <*Amer. Nat.*, 1871, v. 5, pp. 417-419.

87.

1871. PACKARD, ALPHEUS S. [Review of] Scudder and Burgess: Asymmetry in the appendages of hexapod Insects. <*Amer. Nat.*, 1871, v. 5, pp. 420-421.

88.

1871. PACKARD, ALPHEUS S. The embryology of *Chrysopa* and its bearings on the classification of the *Neuroptera*. <*Amer. Nat.*, 1871, v. 5, pp. 564-568.

89.

1871. PACKARD, ALPHEUS S. [Review of] Murray: Geographical distribution of Beetles. <*Amer. Nat.*, 1871, v. 5, pp. 644-646.

90.

1871. PACKARD, ALPHEUS S. [Review of] McLachlan: Position of the Caddis-flies. <*Amer. Nat.*, 1871, v. 5, pp. 707-713.

91.

1871. PACKARD, ALPHEUS S. On the Crustaceans and Insects [of the Mammoth Cave.] <*Amer. Nat.*, 1871, v. 5, pp. 744-761, figs. (See No. 115).

92.

1871. PACKARD, ALPHEUS S. [Fossil Insects, &c., from Sunderland, Mass.] <*Bull. Essex Inst.*, 1871, v. 3, pp. 1-2.

93.

1871. PACKARD, ALPHEUS S. [Abdominal appendages in *Chrysopila* and palpal sacs in *Perla*.] <*Bull. Essex Inst.*, 1871, v. 3, p. 2.

94.

1871. PACKARD, ALPHEUS S. Embryological studies on *Diplax*, *Perithemis*, and the thysanurous genus *Isotoma*. <*Mem. Peab. Acad. Sci.*, 1871, v. 1, No. 2, pp. 24, pl. 1-3, figs.

95.

1871. PACKARD, ALPHEUS S. On the embryology of *Limulus polyphemus*. <*Proc. Amer. Assoc. Adv. Sci.*, 1871, v. 19, pp. 247-255, figs.

96.

1871. PACKARD, ALPHEUS S. Catalogue of the *Phalænidæ* of California. <*Proc. Bost. Soc. Nat. Hist.*, 1871, v. 13, pp. 381-405.

97.

1871. PACKARD, ALPHEUS S. New or rare American *Neuroptera*, *Thysanura*, and *Myriapoda*. <*Proc. Bost. Soc. Nat. Hist.*, 1871, v. 13, pp. 405-411, figs.

98.

1871. PACKARD, ALPHEUS S. Embryology of *Isotoma*, a genus of *Poduridæ*. <*Proc. Bost. Soc. Nat. Hist.*, 1871, v. 14, pp. 13-15, figs.

99.

1871. PACKARD, ALPHEUS S. [Development of *Limulus* and remarks upon the ancestry of Insects.] <*Proc. Bost. Soc. Nat. Hist.*, 1871, v. 14, pp. 60–61.

100.

1871. PACKARD, ALPHEUS S. First annual report on the injurious and beneficial Insects of Massachusetts. <18th *Ann. Rept. Sec. Mass. Bd. Agric.*, 1871, pp. 351–379, pl. 1, figs. Separate: Boston: 1871, pp. 31, pl. 1, figs. (See *Amer. Nat.*, 1871, v. 5, pp. 423–427, figs.)

101.

1871. PACKARD, ALPHEUS S. Report on the *Articulata* [Peab. Acad. Sci.]. <2d and 3d *Ann. Repts. Trustees Peab. Acad. Sci.*, 1871, pp. 62–63.

102.

1871. PACKARD, ALPHEUS S. List of Insects collected at Pebas, Equador, and presented by Prof. James Orton. <2d and 3d *Ann. Rept. Trustees Peab. Acad. Sci.*, 1871, pp. 85–87.

103.

1871. PACKARD, ALPHEUS S. Record of American Entomology for the year 1870. <*Salem*, 1571, pp. 27.

104.

1872. PACKARD, ALPHEUS S. [Review of] Riley: Third Missouri Report. <*Amer. Nat.*, 1872, v. 6, pp. 292–295, figs.

105.

1872. PACKARD, ALPHEUS S. [Review of] Stretch: Illustrations of *Zygænidæ* and *Bombycidæ*. <*Amer. Nat.*, 1872, v. 6, pp. 762–764.

106.

1872. PACKARD, ALPHEUS S. Parthenogenesis in Bees. <*Ann. Bee Cult.*, 1872.

107.

1872. PACKARD, ALPHEUS S. Injurious Insects in Essex County. <*Bull. Essex Inst.*, 1872, v. 4, pp. 5–9, figs.

108.

1872. PACKARD, ALPHEUS S. How many times does the larva of *Arctia caja* change its skin? <*Ent. Mo. Mag.*, 1872, v. 8, p. 206.

109.

1872. PACKARD, ALPHEUS S. On the development of *Limulus polyphemus*. <*Mem. Bost. Soc. Nat. Hist.*, 1872, v. 2, pp. 155–202, pl. 3–5, figs.

110.

1872. PACKARD, ALPHEUS S. Embryological studies on hexapodous Insects. <*Mem. Peab. Acad. Sci.*, 1872, v. 1, No. 3, p. 18, pl. 1–3.

111.

1872. PACKARD, ALPHEUS S. Second annual report on the injurious and beneficial Insects of Massachusetts. <19th Ann. Rept. Sec. Mass. Bd. Agric., 1872, pp. 331-347, figs. Separate: Boston: 1872, pp. 19, figs. (See Amer. Nat., 1873, v. 7, p. 241-244, figs.)

112.

1872. PACKARD, ALPHEUS S. New American Moths: Zygænidæ and Bombycidæ. <4th Ann. Rept. Trustees Peab. Acad. Sci., 1872, p. 84-91. Separate: pp. 8.

113.

1872. PACKARD, ALPHEUS S. List of the Coleoptera collected in Labrador. <4th Ann. Rept. Trustees Peab. Acad. Sci., 1872, pp. 92-94.

114.

1872. PACKARD, ALPHEUS S. Record of American Entomology for the year 1871. <4th Ann. Rept. Trustees Peab. Acad. Sci., 1872, pp. 99-147.

115.

1872. PACKARD, ALPHEUS S. The Mammoth Cave and its inhabitants. <Salem, 1872, pp. 62, figs.
(Dr. Packard contributes Chapter II, on the Crustacea and Insecta. Same as No. 91, with short additional note.)

116.

1873. PACKARD, ALPHEUS S. When is sex determined? <Amer. Nat., 1873, v. 7, pp. 175-177.

117.

1873. PACKARD, ALPHEUS S. On the distribution of Californian Moths. <Amer. Nat., 1873, v. 7, pp. 453-458.

118.

1873. PACKARD, ALPHEUS S. [Review of] Riley: Fifth Missouri Report. <Amer. Nat., 1873, v. 7, pp. 471-477, figs.

119.

1873. PACKARD, ALPHEUS S. Embryology of the Lepidoptera. <Amer. Nat., 1873, v. 7, pp. 486-487.

120.

1873. PACKARD, ALPHEUS S. Farther observations on the embryology of Limulus, with notes on its affinities. <Amer. Nat., 1873, v. 7, pp. 675-678.

121.

1873. PACKARD, ALPHEUS S. Discovery of a Tardigrade. <Amer. Nat., 1873, v. 7, pp. 740-741, figs.

122.

1873. PACKARD, ALPHEUS S. Catalogue of the Pyralidæ of California, with descriptions of new Californian Pterophoridæ. <Ann. Lyc. Nat. Hist. N. Y., 1873, v. 10, pp. 257-267.

123.

1873. PACKARD, ALPHEUS S. Notes on some *Pyralidæ* from New England, with remarks on the Labrador species of the family. <*Ann. Lyc. Nat. Hist. N. Y.*, 1873, v. 10, pp. 267-271.

124.

1873. PACKARD, ALPHEUS S. Catalogue of the *Phalænidæ* of California, No. 2. <*Proc. Bost. Soc. Nat. Hist.*, 1873, v. 16, pp. 13-40, pl. 1.

125.

1873. PACKARD, ALPHEUS S. Occurrence of new and rare Myriapods in Massachusetts. <*Proc. Bost. Soc. Nat. Hist.*, 1873, v. 16, p. 111.

126.

1873. PACKARD, ALPHEUS S. Report of the Curator of Articulates [Peab. Acad. Sci.] <*5th Ann. Rept. Trustees Peab. Acad. Sci.*, 1873, pp. 15-17.

127.

1873. PACKARD, ALPHEUS S. Synopsis of the *Thysanura* of Essex County, Mass., with descriptions of a few extralimital forms. <*5th Ann. Rept. Peab. Acad. Sci.*, 1873, pp. 23-51.

128.

1873. PACKARD, ALPHEUS S. Descriptions of new American *Phalænidæ*. <*5th Ann. Rept. Trustees Peab. Acad. Sci*, 1873, pp. 52-81.

129.

1873. PACKARD, ALPHEUS S. Notes on North American Moths of the families *Phalænidæ* and *Pyralidæ* in the British Museum. <*5th Ann. Rept. Trustees Peab. Acad. Sci.*, 1873, pp. 82-92.

130.

1873. PACKARD, ALPHEUS S. On the cave fauna of Indiana. <*5th Ann. Rept. Trustees Peab. Acad. Sci.*, 1873, pp. 93-97.

131.

1873. PACKARD, ALPHEUS S. Record of American Entomology for the year 1872. <*5th Ann. Rept. Trustees Peab. Acad. Sci.*, 1873, pp. 99-135.

132.

1873. PACKARD, ALPHEUS S. Third annual report on the injurious and beneficial effects of Insects. <*20th Ann. Rept. Sec. Mass. Bd. Agric.*, 1873, p. 237-265, figs. (Reprinted with corrections in *Amer. Nat.*, 1873, v. 7, p. 524-548, figs.)

133.

1873. PACKARD, ALPHEUS S. *Insecta* [of Vineyard Sound.] <*Rept. U. S. Comm. Fish and Fisheries*, 1873, Pt. 1, pp. 539-544.

851—Bull. 16——2

134.

1873. PACKARD, ALPHEUS. S. Descriptions of new species of *Mallophaga* collected by C. H. Merriam while in the government geological survey of the Rocky Mountains, Prof. F. V. Hayden, United States geologist. <*Rept. U. S. Geol. Surv. for* 1872, 1873, pp. 731–734, figs.

135.

1873. PACKARD, ALPHEUS S. Description of new Insects. <*Rept. U. S. Geol. Surv. for* 1872, 1873, pp. 739–741, figs.

136.

1873. PACKARD, ALPHEUS S. Insects inhabiting Great Salt Lake and other saline or alkaline lakes in the West. <*Rept. U. S. Geol. Surv. for* 1872, 1873, pp. 743–746.

137.

1873. PACKARD, ALPHEUS S. Directions for collecting and preserving Insects, prepared for the use of the Smithsonian Institution. <*Smith. Misc. Coll.*, 1873, v. 11, pp. 3+55, figs. Separate: Washington, 1873, pp. 3+55, figs.

138.

1873. PACKARD, ALPHEUS S. On the ancestry of Insects. <*Salem*, 1873. (Printed in advance from Our Common Insects.)

139.

1873. PACKARD, ALPHEUS S. Our Common Insects. A popular account of the Insects of our fields, forests, gardens, and houses. <*Salem*, 1873, pp. 16+225, pl., figs.

140.

1874. PACKARD, ALPHEUS S. [Morphology of Insects; reply to criticism of C. V. Riley.] <*Amer. Nat.*, 1874, v. 8, pp. 187–189].

141.

1874. PACKARD, ALPHEUS S. Occurrence of *Telea polyphemus* in California. A correction. <*Amer. Nat.*, 1874, v. 8, pp. 243–244.

142.

1874. PACKARD, ALPHEUS S. Nature's means of limiting the numbers of Insects. <*Amer. Nat.*, 1874, v. 8, pp. 270–282, figs.

143.

1874. PACKARD, ALPHEUS S. The discovery of the origin of the sting of the Bee. <*Amer. Nat.*, 1874, v. 8, p. 431.

144.

1874. PACKARD, ALPHEUS S. The mouth-parts of the Dragon-fly. <*Amer. Nat.*, 1874, v. 8, p. 432.

145.

1874. PACKARD, ALPHEUS S. Occurrence of *Japyx* in the United States. <*Amer. Nat.*, 1874, v. 8, pp. 501–502, fig.

19

146.

1874. PACKARD, ALPHEUS S. The "hateful" Grasshopper in New England. $<$ *Amer. Nat.*, 1874, v. 8, p. 502.

147.

1874. PACKARD, ALPHEUS S. [Grasshoppers as food.] $<$ *Amer. Nat.*, 1874, v. 8, p. 511.

148.

1874. PACKARD, ALPHEUS S. On the distribution and primitive number of spiracles in Insects. $<$ *Amer. Nat.*, 1874, v. 8, pp. 531–534.

149.

1874. PACKARD, ALPHEUS S. Larvæ of *Anophthalmus* and *Adelops*. $<$ *Amer. Nat.*, 1874, v. 8, pp. 562–563.

150.

1874. PACKARD, ALPHEUS S. The metamorphosis of Flies. 1–III. $<$ *Amer. Nat.*, 1874, v. 8, pp. 603–612; 661–667; 713–721. (*Translated from A. Weismann.*)

151.

1874. PACKARD, ALPHEUS S. Further observations on the embryology of *Limulus*, with notes on its affinities. $<$ *Proc. Amer. Assoc. Adv. Sci.*, 1874, v. 22, pp. 30–32.

152.

1874. PACKARD, ALPHEUS S. On the transformations of the common House-fly, with notes on allied forms. $<$ *Proc. Bost. Soc. Nat. Hist.*, 1874, v. 16, pp. 136–150, pl. 3, figs.

153.

1874. PACKARD, ALPHEUS S. Report of the Curator of *Articulata* [Peab. Acad. Sci.] $<$ *6th Ann. Rept. Trustees Peab. Acad. Sci.*, 1874, pp. 13–14.

154.

1874. PACKARD, ALPHEUS S. Descriptions of new North American *Phalænidæ*. $<$ *6th Ann. Rept. Trustees Peab. Acad. Sci.*, 1874, pp. 39–53.

155.

1874. PACKARD, ALPHEUS S. Record of American Entomology for the year 1873. $<$ *6th Ann. Rept. Trustees Peab. Acad. Sci.*, 1874, pp. 61–114.

156.

1874. PACKARD, ALPHEUS S. [Parasites of White Mountain Butterflies.] $<$ *Final Rept. Geol. N. H.*, 1874, v. 1, p. 347, figs.

157.

1874. PACKARD, ALPHEUS S. On the geographical distribution of the Moths of Colorado. $<$ *Rept. U. S. Geol. Surv. for 1873*, 1874, pp. 543–560, figs.

158.

1874. PACKARD, ALPHEUS S. Report on the Myriopods collected by Lieut. W. L. Carpenter in 1873 in Colorado. $<$ *Rept. U. S. Geol. Surv. for 1873*, 1874, p. 607.

20

159.

1875. PACKARD, ALPHEUS S. [Proposed monograph of] the Geometrid Moths. <*Amer. Nat.*, 1875, v. 9, pp. 64, 179-180, figs.

160.

1875. PACKARD, ALPHEUS S. The Invertebrate cave fauna of Kentucky and adjoining states. *Araneina.* <*Amer. Nat.*, 1875, v. 9, pp. 274-278.

161.

1875. PACKARD, ALPHEUS S. [Cigars destroyed by Insects.] <*Amer. Nat.*, 1875, v. 9, p. 375.

162.

1875. PACKARD, ALPHEUS S. On the development of the nervous system in *Limulus.* <*Amer. Nat.*, 1875, v. 9, pp. 422-424.

163.

1875. PACKARD, ALPHEUS S. On an undescribed organ in *Limulus*, supposed to be renal in its nature. <*Amer. Nat.*, 1875, v. 9, pp. 511-514; *Ann. and Mag. Nat. Hist.*, 1875, ser. 4, v. 15, pp. 255-258.

164.

1875. PACKARD, ALPHEUS S. *Calopteuus spretus* [= *C. atlanis*] in Massachusetts. <*Amer. Nat.*, 1875, v. 9, p. 573.

165.

1875. PACKARD, ALPHEUS S. Life histories of the *Crustacea* and Insects. <*Amer. Nat.*, 1875, v. 9, pp. 583-622, figs.

166.

1875. PACKARD, ALPHEUS S. Cave-inhabiting Spiders. <*Amer. Nat.*, 1875, v. 9, pp. 663-664.

167.

1875. PACKARD, ALPHEUS S. On gynandromorphism in the *Lepidoptera*. <*Mem. Bost. Soc. Nat. Hist.*, 1875, v. 2, pp. 409-412, pl. 14, in part.

168.

1876. PACKARD, ALPHEUS S. The cave Beetles of Kentucky. <*Amer. Nat.*, 1876, v. 10, pp. 282-287, pl. 2, figs.

169.

1876. PACKARD, ALPHEUS S. The House-fly. <*Amer. Nat.*, 1876, v. 10, pp. 476-480, figs.

170.

1876. PACKARD, ALPHEUS S. [Review of] Riley: Eighth Missouri Report. <*Amer. Nat.*, 1876, v. 10, pp. 485-486.

171.

1876. PACKARD, ALPHEUS S. A century's progress in American Zoology. <*Amer. Nat.*, 1876, v. 10, pp. 591–598. *Gerrais Journ. de Zool.*, 1876, v. 5, pp. 413–423.

172.

1876. PACKARD, ALPHEUS S. [Review of] Cook: Manual of the Apiary. <*Amer. Nat.*, 1876, v. 10, pp. 621–622.

173.

1876. PACKARD, ALPHEUS S. [Review of] Mayer: Ontogeny and phylogeny of Insects. <*Amer. Nat.*, 1876, v. 10, pp. 688–691.

174.

1876. PACKARD, ALPHEUS S. [The ravages of Locusts.] <*Amer. Nat.*, 1876, v. 10, pp. 754–755.

175.

1876–1878. PACKARD, ALPHEUS S. Johnson's New Universal Cyclopædia. <*New York*, 1876–1878.
Dr. PACKARD contributes:
1. *Hymenoptera*, 1876, v. 2, pp. 1075–1076.
2. *Lepidoptera* .... pp. 1733–1734.
3. *Locust*, 1877 .... v. 3, p. 87.
4. *Louse* .... p. 129.
5. *Neuroptera* .... pp. 782–783.
6. *Silk-worm*, 1878 ... v. 4, pp. 1662–1664.

176.

1876. PACKARD, ALPHEUS S. A monograph of the Geometrid Moths or *Phalænidæ* of the United States. <*Rept. U. S. Geol. Surv., Washington*, 1876, pp. 4+607, pl. 1–13, figs.

177.

1876. PACKARD, ALPHEUS S. Life histories of animals, including man, or outlines of comparative embryology. <*New York*, 1876, pp. 243, pl., figs.

178.

1877. PACKARD. ALPHEUS S. The migrations of the destructive Locust of the West. <*Amer. Nat.*, 1877, v. 11, pp. 22–29.

179.

1877. PACKARD, ALPHEUS S. Explorations of the Polaris expedition to the North Pole. <*Amer. Nat.*, 1877, v. 11, pp. 51–53. Separate, 1877, pp. 2. (See *Ent. Mo. Mag.*, 1877, v. 13, pp. 228–229.)

180.

1877. PACKARD, ALPHEUS S. Partiality of white Butterflies for white flowers. <*Amer. Nat.*, 1877, v. 11, p. 243.

181.

1877. PACKARD, ALPHEUS S. Experiments on the sense organs of Insects. <*Amer. Nat.*, 1877, v. 11, pp. 418–423.

22

182.

1377. PACKARD, ALPHEUS S. [Review of] Murray: Aptera. <*Amer. Nat.*, 1877, v. 11, pp. 482–4–3.

183.

1877. PACKARD, ALPHEUS S. United States Entomological Commission:* Circular No. 1. [Riley, Packard, Thomas.] <*Washington:* 1877, pp. 4.

184.

1877. PACKARD, ALPHEUS S. United States Entomological Commission : Circular No. 3. <*Washington:* 1877.

185.

1877. PACKARD, ALPHEUS S. United States Entomological Commission : Bulletin No. 1. [Riley, Packard, Thomas.] <*Washington:* 1877, pp. 12.

186.

1877. PACKARD, ALPHEUS S. United States Entomological Commission : Bulletin No. 2. [Riley, Packard, Thomas.] <*Washington:* 1877, pp. 14, figs.

187.

1877. PACKARD, ALPHEUS S. On a new cave fauna in Utah. <*Bull. U. S. Geol. and Geogr. Surv.*, 1877, v. 3, pp. 157–169, figs.

188.

1877. PACKARD, ALPHEUS S. The Hessian-fly, Joint-worm, and Wheat-midge. <*Ca. Ent.*, 1877, v. 9, p. 100.

189.

1877. PACKARD, ALPHEUS S. Experiments upon the vitality of Insects. <*Psyche*, 1877, v. 2, pp. 17–19.

190.

1877. PACKARD, ALPHEUS S. Appendages homologous with legs. <*Psyche*, 1877, v. 2, p. 23.

191.

1877. PACKARD, ALPHEUS S. Report on the Rocky Mountain Locust and other Insects now injuring or likely to injure field and garden crops in the western states and territories. <*Rept. U. S. Geol. Surv. for* 1875, 1877, pp. 589–810, pls. 62–70, maps 1–5, figs.

192.

1877. PACKARD, ALPHEUS S. List of *Coleoptera* collected in 1875 in Colorado and Utah by A. S. Packard, jr., M. D. <*Rept. U. S. Geol. Surv. for* 1875, 1877, pp. 811–815.

193.

1877. PACKARD, ALPHEUS S. Half-Hours with Insects. <Boston : 1877, pp. 8+384, pl., figs.

---

*The publications of the Commission were prepared unitedly by Messrs. Riley, Packard, and Thomas, but to the individual members were assigned special work and chapters in the subdivision of the work.

194.

1878. PACKARD, ALPHEUS S. The mode of extrication of Silk-worm Moths from their cocoons. <*Amer. Nat.*, 1878, v. 12, pp. 379–383, figs; *Nature*, 1878, v. 18, pp. 226–227, figs; *Ent. Nach.*, 1878, v. 5, pp. 284–285.

195.

1878. PACKARD, ALPHEUS S. Some characteristics of the central zoo-geographical province of the United States. <*Amer. Nat.*, 1878, v. 12, pp. 512–517.

196.

1878. PACKARD, ALPHEUS S. Insects affecting the Cranberry, with remarks on other injurious Insects. <*Rept. U. S. Geol. Surv. for* 1876, 1878, pp. 521–531, figs.

197.

1878. PACKARD, ALPHEUS S. First annual report of the U. S. Entomological Commission. <*Washington*, 1878, pp. 14+477+295, pl. 1–5, figs.
    Dr. Packard contributes:
        Chapter 2. Chronological history, pp. 53–113.
        Chapter 5. Permanent breeding grounds of the Rocky Mountain Locust, pp. 131–136.
        Chapter 9. Anatomy and embryology, pp. 257–279.
        Chapter 17. Uses to which Locusts may be put, pp. 437–443.
        Chapter 19. Locust ravages in other countries, pp. 460–477.
        Appendix IX. Narrative of the first journey made in the summer of 1877, pp. 134–138.
        Appendix X. Narrative of a second journey made in the summer of 1877, pp. 139–144.
    With the co-operation of Mr. Cyrus Thomas, Dr. Packard contributes:
        Chapter 6. Geographical distribution, pp. 136–142.
        Chapter 7. Migrations, pp. 143–211.

198.

1878. PACKARD, ALPHEUS S. Insects of the West: An account of the Rocky Mountain Locust, the Colorado Potato-beetle, the Canker-worm, Currant Saw-fly, and other Insects which devastate the crops of the country. <*London*, 1878. (A reprint, with slight changes, paging, &c., of No. 191.)

199.

1879. PACKARD, ALPHEUS S. [The smallest Insects known.] <*Amer. Nat.*, 1879, v. 13, p. 62.

200.

1879. PACKARD, ALPHEUS S. [Fossil Insects at Green River, Wyoming.] <*Amer. Nat.*, 1879, v. 13, p. 203.

201.

1879. PACKARD, ALPHEUS S. [Parthenogenesis of the Honey-bee.] <*Amer. Nat.*, 1879, v. 13, p. 394.

202.

1879. PACKARD, ALPHEUS S. A poisonous Centipede. <*Amer. Nat.*, 1879, v. 13, p. 527.

203.

1879. PACKARD, ALPHEUS S. [Cotton-worm investigation.] <*Amer. Nat.*, 1879, v. 13, p. 535.

204.

1879. PACKARD, ALPHEUS S. The Rocky Mountain Locust in New Mexico. <*Amer. Nat.*, 1879, v. 13, p. 526.

205.

1879. PACKARD, ALPHEUS S. [Review of] Graber: *Die Insekten.* <*Amer. Nat.*, 1879, v. 13, pp. 774–775.

206.

1879. PACKARD, ALPHEUS S. Zoology for students and general readers. <*New York*: 1879, pp. 8+719, figs.
  a. 2d edition, New York, 1880.
  b. 3d edition, New York, 1881.
  c. 4th edition, New York, 1883.
  d. 5th edition, New York, 1886.

207.

1879. PACKARD, ALPHEUS, S. Zoology of the Invertebrate animals. <*New York :* 1879, pp. 12+143, figs. (By A. Macalister. Specially revised for America by A. S. Packard.)

208.

1880. PACKARD, ALPHEUS S. Moths entrapped by an Asclepiad plant, Physianthus, and killed by Honey-bees. <*Amer. Nat.*, 1880, v. 14, pp. 48–50; *Nature*, 1880, v. 21, p. 308; *Journ. Roy. Micros. Soc.*, 1880, ser 1, v. 3, pp. 241–242; *Journ. Sci.*, 1880, ser. 3, v. 2, p. 213; *Bot. Gaz.*, 1880, v. 5, pp. 17–20.

209.

1880. PACKARD, ALPHEUS S. The Cotton-worm Moth in Rhode Island. <*Amer. Nat.*, 1880, v. 14, p. 53.

210.

1880. PACKARD, ALPHEUS S. Structure of the eye of *Limulus.* <*Amer. Nat.*, 1880, v. 14, pp. 212–213; *Ann. and Mag., Nat. Hist.*, 1880, ser. 5, v. 5, pp. 434–435; *Journ. Roy. Micros. Soc.*, 1880, ser. 1, v. 3, pp. 947–948.

211.

1880. PACKARD, ALPHEUS S. On the internal structure of the brain of *Limulus polyphemus.* <*Amer. Nat.*, 1880, v. 14, pp. 445–448; *Ann. and Mag. Nat. Hist.* 1880, ser. 5, v. 6, pp. 29–33

212.

1880. PACKARD, ALPHEUS S. Case of protective mimicry in a Moth. <*Amer. Nat.*, 1880, v. 14, p. 600.

213.

1880. PACKARD. ALPHEUS S. The eyes and brain of *Cermatia forceps.* <*Amer. Nat.*, 1880, v. 14, pp. 602–603. *Journ. Roy. Micros. Soc.*, 1880, ser. 1, v. 3, pp. 783–784.

214.

1880. PACKARD, ALPHEUS S. [Review of] Wood-Mason: Morphology of Insects. <*Amer. Nat.*, 1880, v. 14, pp. 665–667.

215.

1880. PACKARD, ALPHEUS S. [The investigations of the U. S. Entomological Commission for 1880.] <*Amer. Nat.*, 1880, v. 14, pp. 753–755.

216.

1880. PACKARD, ALPHEUS S. Eggs of Tree-crickets wanted. <*Amer. Nat.*, 1880, v. 14, p. 804.

217.

1880. PACKARD, ALPHEUS S. *Cetonia inda.* <*Amer. Nat.*, 1880, v. 14, p. 806.

218.

1880. PACKARD, ALPHEUS S. The anatomy, histology, and embryology of *Limulus polyphemus.* <*Annis Mem. Bost. Soc. Nat. Hist.*, 1880, pp. 45, pls. 1–7 ; *Journ. Roy. Micros. Soc.*, 1881, ser. 2, v. 1, pp. 600–601.

219.

1880. PACKARD, ALPHEUS S. The Hessian fly, its ravages, habits, enemies, and means of preventing its increase. <*Bull. U. S. Ent. Comm., No.* 4, 1880, pp. 43, pls. 1–2, map, fig. (See *Amer. Nat.*, 1880, v, 14, pp. 586–587 ; *Amer. Ent.*, 1880, v. 3, pp. 118–121, 140–141, figs.)

220.

1880. PACKARD, ALPHEUS S. On the internal structure of the brain of *Limulus polyphemus.* <*Zool. Anz.*, 1880, v. 3, pp. 306–310.

221.

1880. PACKARD, ALPHEUS S. Second report of the U. S. Entomological Commission. <*Washington*, 1880, pp. 18+322+80, pls. 1–17, maps, figs.

  Dr. PACKARD contributes :

    Chapter 6. The southern limits of the distribution of the Rocky Mountain Locust, pp. 156–160.

    Chapter 7. Summer of Locust flights from 1877 to 1879, pp. 160–163.

    Chapter 8. The Western Cricket, pp. 163–178.

    Chapter 9. The air-sacs of Locusts, with reference to their powers of flight, pp. 178–183.

    Chapter 11. The brain of the Locust, pp. 223–242.

    Appendix VII. Notes of a journey made to Utah and Idaho in the summer of 1878, pp. 69–71.

    Appendix VIII. Yersin's researches on the functions of the nervous system of the Articulate animals, pp. 73–74.

  With the co-operation of Professor RILEY, Dr. PACKARD contributes :

  Chapter 1. Additions to the chronology of Locust ravages, pp. 1–14.

222.

1881. PACKARD, ALPHEUS S. Fauna of the Luray and the Newmarket Caves, Virginia. <*Amer. Nat.*, 1881, v. 15, pp. 231–232.

223.

1881. PACKARD, ALPHEUS S. The brain of the Locust. <*Amer., Nat.* 1881, v. 15, pp. 285–302, pl. 1–3; *Journ. de Microg.*, 1881, v. 5, pp. 448–453; v. 6, pp. 71–75.

224.

1881. PACKARD, ALPHEUS S. The brain of the embryo and young Locust. <*Amer. Nat.*, 1881, v. 15, pp. 372–379, pl. 4–5.

225.

1881. PACKARD, ALPHEUS S. Locusts in Mexico in 1880. <*Amer. Nat.* 1881, v. 15, p. 578.

226.

1881. PACKARD, ALPHEUS S. *Scolopendrella* and its position in nature. <*Amer. Nat.*, 1881, v. 15, pp. 698–704, figs.

227.

1881. PACKARD, ALPHEUS S. The fauna of Nickajack Cave. <*Amer. Nat.*, 1881, v., 15, pp. 877–882, pl. 7. (Prepared by Prof. E. D. Cope and Dr. Packard.)

228.

1881. PACKARD, ALPHEUS S. [Review of] Scudder: Butterflies. <*Amer. Nat.*, 1881, v. 15, pp. 885–887.

229.

1881. PACKARD, ALPHEUS S. Insects injurious to forest and shade trees. <*Bull. U. S. Ent. Comm.*, No. 7, 1881, pp. 275, figs.

230.

1881. PACKARD, ALPHEUS S. Bibliography of economic Entomology. <*Ca. Ent.*, 1881, v, 13, p. 39. (Plan of proposed work.)

231.

1881. PACKARD, ALPHEUS S. Descriptions of some new Ichneumon parasites of North American Butterflies. <*Proc. Bost. Soc. Nat. Hist.*, 1881, v. 21, pp. 18–38.

232.

1881. PACKARD, ALPHEUS S. The Grasshopper question. <*Rocky Mt. Husb.*, 1881.

233.

1882. PACKARD, ALPHEUS S. [Review of] Thomas: Fifth Illinois Report. <*Amer. Nat.*, 1882, v. 16, pp. 39–40.

234.

1882. PACKARD, ALPHEUS S. Is *Limulus* an Arachnid? <*Amer. Nat.*, 1882, v. 16, pp. 287–292; *Ann. and Mag. Nat. Hist.*, 1882, ser. 5, v. 9, pp. 369–374; *Journ. Roy. Micros. Soc.*, 1882, ser. 2, v. 2, p. 337.

235.

1882. PACKARD, ALPHEUS S. A correction [to No. 234]. <*Amer. Nat.*, 1882, v. 16, p. 436.

236.

1882. PACKARD, ALPHEUS S. The coloring of zoo-geographical maps. <*Amer. Nat.*, 1882, v. 16, p. 589.

237.

1882. PACKARD, ALPHEUS S. Bot-fly maggots in a Turtle's neck. <*Amer. Nat.*, 1882, vol. 16, p. 598, figs.

238.

1882. PACKARD, ALPHEUS S. Larvæ of a Fly in a Hot Spring in Colorado. <*Amer. Nat.*, 1882, v. 16, pp. 599–600.

239.

1882. PACKARD, ALPHEUS S. Nomenclature of external parts of *Arthropoda*. <*Amer. Nat.*, 1882, v. 16, pp. 676–677.

240.

1882. PACKARD, ALPHEUS S. [Review of] Lubbock : Ants, bees, and wasps. <*Amer. Nat.*, 1882, v. 16, pp. 804–807.

241.

1882. PACKARD, ALPHEUS S. Probable difference in two broods of *Drasteria erechthea*. <*Papilio*, 1882, v. 2, pp. 147–148.

242.

1882. PACKARD, ALPHEUS S. Notes on lepidopterous larvæ. <*Papilio*, 1882, v. 2, pp. 180–183.

243.

1883. PACKARD, ALPHEUS S. The systematic position of the *Archipolypoda*, a group of fossil Myriopods. <*Amer. Nat.*, 1883, v. 17, pp. 326–329, figs ; *Journ. Roy. Micros. Soc.*, 1883, ser. 2, v. 3, pp. 365–365.

244.

1883. PACKARD, ALPHEUS S. [Review of] Riley : Rept. U. S. Ent. for 1881–1882. <*Amer. Nat.*, 1883, v. 17, pp. 399–400, figs.

245.

1883. PACKARD, ALPHEUS S. A new species of *Polydesmus* with eyes. <*Amer. Nat.*, 1883, v. 17, pp. 428–429, figs. ; *Journ. Roy. Micros. Soc.*, 1883, ser. 2, v. 3, p. 367.

246.

1883. PACKARD, ALPHEUS S. Discovery of *Eurypauropus* in Europe. <*Amer. Nat.*, 1883, v. 17, p. 555.

247.

1883. PACKARD, ALPHEUS S. Repugnatorial pores in the *Lysiopetalidæ*. <*Amer. Nat.*, 1883, v. 17, p. 555.

248.

1883. PACKARD, ALPHEUS S. [Review of] Recent works on the mouth-parts of Flies. <*Amer. Nat.*, 1883, v. 17, pp. 631–633.

249.

1883. PACKARD, ALPHEUS S. The coxal gland of Arachnids and *Crustacea*. <*Amer. Nat.*, 1883, v. 17, pp. 795–797.

250.

1883. PACKARD, ALPHEUS S. On the classification of the Linnæan orders of *Orthoptera* and *Neuroptera*. <*Amer. Nat.*, 1883, v. 17, pp. 820–829; *Ann. and Mag. Nat. Hist.*, 1883, ser. 5, v. 12, pp. 145–154; *Journ. Roy. Micros. Soc.*, 1884, ser. 2, v. 4, p. 220.

251.

1883. PACKARD, ALPHEUS S. Note on a *Peripatus* from the Isthmus of Panama. <*Amer. Nat.*, 1883, v. 17, pp. 881–882, figs.

252.

1883. PACKARD, ALPHEUS S. The structure and embryology of *Peripatus*. <*Amer. Nat.*, 1883, v. 17, pp. 882–884.

253.

1883. PACKARD, ALPHEUS S. On the genealogy of the Insects. <*Amer. Nat.*, 1883, v. 17, pp. 932–945, figs.; *Journ. Roy. Micros. Soc.*, 1884, ser. 2, v. 4, pp. 217–218; *Journ. de Microg.*, 1884, v. 7, pp. 566–571, 622–628, figs; *Bull. Soc. Ent. Ital.*, 1884, v. 16, pp. 135–136.

254.

1883. PACKARD, ALPHEUS S. [Review of] Weismann: Studies in the theory of descent. <*Amer. Nat.*, 1883, v. 17, pp. 1042–1046.

255.

1883. PACKARD, ALPHEUS S. Molting in the shell in *Limulus*. <*Amer. Nat.*, 1883, v. 17, pp. 1075–1076; *Journ. Roy. Micros. Soc.*, 1883, ser. 2, v. 3, pp. 836–837.

256.

1883. PACKARD, ALPHEUS S. The number of segments in the head of winged Insects. <*Amer. Nat.*, 1883, v. 17, pp. 1134–1138, figs.; *Journ. Roy. Micros. Soc.*, 1884, ser. 2, v. 4, pp. 43–44.

257.

1883. PACKARD, ALPHEUS S. Occurrence of a *Stratiomys* larva in sea-water. <*Amer. Nat.*, 1883, v. 17, pp. 1287–1288.

258.

1883. PACKARD, ALPHEUS S. Note on forest-tree Insects. <*Bull. Div. Ent. U. S. Dept. Agric.*, No. 3, 1883, pp. 24–30.

259.

1883. PACKARD, ALPHEUS S. Scorpions. <*Independent*, 1883.

260.

1883. PACKARD, ALPHEUS S. Decay of the spruce in the Adirondacks and northern New England. <*Nation*, 1883, v. 37, p. 525.

261.

1883. PACKARD, ALPHEUS S. A revision of the *Lysiopetalidæ*, a family of Chilognath *Myriopoda*, with a notice of the genus *Cambala*. <*Proc. Amer. Philos. Soc.*, 1883, v. 21, pp. 177–197 ; *Journ. Roy. Micros. Soc.*, 1883, ser. 2, v. 3, p. 832.

262.

1883. PACKARD, ALPHEUS S. On the morphology of the *Myriopoda*. <*Proc. Amer. Philos. Soc.*, 1883, v. 21, pp. 197–209, figs.; *Journ. Roy. Micros. Soc.*, 1883, ser. 2, v. 3, pp. 832–833.

263.

1883. PACKARD, ALPHEUS S. Report on the causes of destruction of evergreen forests in northern New England and New York. <*Rept. Dept. Agric. for 1883, [part of Riley's report as Entomologist]*, 1883, pp. 138–151, pl. 9, figs.

264.

1883. PACKARD, ALPHEUS S. Third report of the U. S. Entomological Commission. <*Washington*, 1883, pp. 14+347+92, pl. 1–64, maps, figs.
> Dr. PACKARD contributes :
> Chapter 8. The Hessian fly, pp. 193–248.
> Chapter 9. Descriptions of the larvæ of injurious forest insects, pp. 251–262.
> Chapter 10. The embryological development of the Locust, pp. 263–285.
> Chapter 11. The systematic position of the *Orthoptera* in relation to the other orders of Insects, pp. 286–345.
> Chapter 12. Note on the geographical distribution of the Rocky Mountain Locust, illustrated with colored zoo-geographical map of North America, pp. 346–347.

265.

1883. PACKARD, ALPHEUS S. The Standard Natural History. <*Boston*, 1883–1885.
> Dr. PACKARD contributes :
> 1. Introduction, 1885, v. 1, pp. i–lxxii.
> 2. *Hexapoda*, 1883, v. 2, pp. 131–134.
> 3. *Thysanura*, pp. 135–138.
> 4. *Dermatoptera*, p. 139.
> 5. *Pseudoneuroptera*, 1883–'84, v. 2, pp. 140–154.
> 6. *Neuroptera*, 1884, v. 2, pp. 155–166.

266.

1883. PACKARD, ALPHEUS S. Zoology. Briefer Course. <*New York*, 1883, pp. 5+334.
> a. 2d edition, New York, 1885.
> b. 3d edition, New York, 1886.

267.

1884. PACKARD, ALPHEUS S. Sograff's embryology of the Chilopod Myriopods. <*Amer. Nat.*, 1884, v. 18, pp. 201–202.

268.

1884. PACKARD, ALPHEUS S. New cave Arachnids. <*Amer. Nat.*, 1884, v. 18, pp. 202-204, figs.

269.

1884. PACKARD, ALPHEUS S. [Review of] Meinert : Caput *Scolopendrœ*. <*Amer. Nat.*, 1884, v. 18, pp. 270-272.

270.

1884. PACKARD, ALPHEUS S. Egg-laying habits of the egg parasite of the Canker-worm. <*Amer. Nat*, 1884, v. 18, pp. 292-293.

271.

1884. PACKARD, ALPHEUS S. Paired sexual outlets in Insects. <*Amer. Nat.*, 1884, v. 18, p. 293.

272.

1884. PACKARD, ALPHEUS S. The Larch-worm. <*Amer. Nat.*, 1884, v. 18, pp. 293-296, figs.

273.

1884. PACKARD, ALPHEUS S. The Hemlock *Gelechia*. <*Amer. Nat.*, 1884. v. 18, p. 296.

274.

1884. PACKARD, ALPHEUS S. The Spruce-bud *Tortrix*. <*Amer. Nat.*, 1884, v. 18, pp. 424-426, figs.

275.

1884. PACKARD, ALPHEUS S. Notes on Moths. <*Amer. Nat.*, 1884, v. 18, pp. 632-633.

276.

1884. PACKARD, ALPHEUS S. The transformations of *Nola*. <*Amer. Nat.*, 1884, v. 18, pp. 726-727.

277.

1884. PACKARD, ALPHEUS S. Habits of an aquatic Pyralid caterpillar. <*Amer. Nat.*, 1884, v. 18, pp. 824-826, pl. 24.

278.

1884. PACKARD, ALPHEUS S. Note on salt-water Insects, No. 3. <*Amer. Nat.*, 1884, v. 18, pp. 826-828, figs.

279.

1884. PACKARD, ALPHEUS S. Aspects of the body in Vertebrates and Arthropods. <*Amer. Nat.*, 1884, v. 18, pp. 855-861, figs. ; *Ann. and Mag. Nat. Hist.*, 1884, ser. 5, v. 14, pp. 243-249, figs. ; *Journ. Roy. Micros. Soc.*, 1884, ser. 2, v. 4, p. 866.

280.

1884. PACKARD, ALPHEUS S. Life-histories of some Geometrid Moths. <*Amer. Nat.*, 1884, v. 18, pp. 933-936.

281.

1884. PACKARD, ALPHEUS S. Anatomy and function of the Bee's tongue. <*Amer. Nat.*, 1884, v. 18, p. 937.

282.

1884. PACKARD, ALPHEUS S. Life-history of *Lochmæus tessella*. <*Amer. Nat.*, 1884, v. 18, pp. 1044-1045.

283.

1884. PACKARD, ALPHEUS S. Transformations of *Caripeta angustiorata*. <*Amer. Nat.*, 1884, v. 18, pp. 1045-1046.

284.

1884. PACKARD, ALPHEUS S. Mode of oviposition of the common Longicorn Pine borer (*Monohammus confusor*). <*Amer. Nat.*, 1884, v. 18, pp. 1149-1151.

285.

1884. PACKARD, ALPHEUS S. Egg-laying habits of the Maple-tree borer. <*Amer. Nat.*, 1884, v. 18, pp. 1151-1152.

286.

1884. PACKARD, ALPHEUS S. Palmen's paired outlets of the sexual organs of Insects. <*Amer. Nat.*, 1884, v. 18, pp. 1152-1153.

287.

1884. PACKARD, ALPHEUS S. The nature of the so-called "liver" in the Arachnids. <*Amer. Nat.*, 1884, v. 18, pp. 1153-1154.

288.

1884. PACKARD, ALPHEUS S. The systematic position of the *Embidæ*. <*Amer. Nat.*, 1884, v. 18, pp. 1154-1155.

289.

1884. PACKARD, ALPHEUS S. The larval stages of *Mamestra picta*. <*Amer. Nat.*, 1884, v. 18, pp. 1266-1267.

290.

1884. PACKARD, ALPHEUS S. The Bees, Wasps, &c., of Labrador. <*Amer. Nat.*, 1884, v. 18, p. 1267.

291.

1884. PACKARD, ALPHEUS S. Origin of Bee's cells. <*Amer. Nat.*, 1884, v. 18, pp. 1268-1269.

292.

1884. PACKARD, ALPHEUS S. The life of the Great Salt Lake. I–IV. <*Independent*, 1884.

293.

1885. PACKARD, ALPHEUS S. [Review of] Dahl: Structure and function of the legs of Insects. <*Amer. Nat.*, 1885, v, 19, pp. 178-180.

294.

1885. PACKARD, ALPHEUS S. The number of abdominal segments in lepidopterous larvæ. <*Amer. Nat.*, 1885, v. 19, pp. 307–308; *Journ. Roy. Micros. Soc.*, 1885, ser. 2, v. 5, p. 636.

295.

1885. PACKARD, ALPHEUS S. [Review of] Riley: Rept. U. S. Ent. for 1884. <*Amer. Nat.*, 1885, v. 19, p. 607, pl. 18.

296.

1885. PACKARD, ALPHEUS S. [Review of] Smith: Systematic position of some N. A. Lepidoptera. <*Amer. Nat.*, 1885, v. 19, pp. 608–609.

297.

1885. PACKARD, ALPHEUS S. Unusual number of legs in the caterpillar of *Lagoa*. <*Amer. Nat.*, 1885, v. 19, pp. 714–715, figs.; *Journ. Roy. Micros. Soc.*, 1885, ser. 2, v. 5, pp. 990–991.

298.

1885. PACKARD, ALPHEUS S. Use of the pupæ of Moths in distinguishing species. <*Amer.Nat.*, 1885, v. 19, pp. 715–716.

299.

1885. PACKARD, ALPHEUS S. On the embryology of *Limulus polyphemus*. III. <*Amer. Nat.*, 1885, v. 19, pp. 722–727, pl. 24 ; *Proc. Amer. Philos. Soc.*, 1885, v. 22, pp. 268–272, pl. ; *Journ. Roy. Micros. Soc.*, 1885, ser. 2, v. 5, pp. 806–807.

300.

1885. PACKARD, ALPHEUS S. Edible Mexican Insects. <*Amer. Nat.*, 1885, v. 19, p. 893.

301.

1885. PACKARD, ALPHEUS S. Dr. Brauer's views on the classification of Insects. <*Amer. Nat.*, 1885, v. 19, pp. 999–1001.

302.

1885. PACKARD, ALPHEUS S. Flights of Locusts in Eastern Mexico in 1885. <*Amer. Nat.*, 1885, v. 19, pp. 1105–1106.

303.

1885. PACKARD, ALPHEUS S. [Review of] Hickson · Eye and optic tract of Insects. <*Amer. Nat.*, 1885, v. 19, pp. 1220–1221.

304.

1885. PACKARD, ALPHEUS S. Spiders. <*Random Notes on Nat. Hist.*, 1885, v. 2, p. 11.

305.

1885. PACKARD, ALPHEUS S. Second report on the causes of the destruction of the evergreen and other forest trees in northern New England and New York. <*Rept. Dept. Agric. for 1884,* [*part of Riley's Report as Entomologist,*] 1885, pp. 374–383, figs. Separate: 1885, pp. 12, figs.

306.

1886. PACKARD, ALPHEUS S. [Review of] Brongniart: Studies of carboniferous Insects. <*Amer. Nat.*, 1886, v. 20, pp. 68–69.

307.

1886. PACKARD, ALPHEUS S. [Review of] Plateau: Experiments on vision of Insects. <*Amer. Nat.*, 1886, v. 20, pp. 69–70.

308.

1886. PACKARD, ALPHEUS S. [Review of] Cholodkoosky: Morphology of *Lepidoptera*. <*Amer. Nat.*, 1886, v. 20. pp. 169–170.

309.

1886. PACKARD, ALPHEUS S. Flights of Locusts at San Luis Potosi, Mexico, 1885. <*Amer. Nat.*, 1886, v. 28, p. 170.

310.

1886. PACKARD, ALPHEUS S. [Review of] Scudder: Systematische Übersicht fossilen Insekten. <*Amer. Nat.*, 1886, v. 20, pp. 369–370.

311.

1886. PACKARD, ALPHEUS S. On the cinurous *Thysanura* and *Symphyla* of Mexico. <*Amer. Nat.*, 1886, v. 20, pp. 382–383.

312.

1886. PACKARD, ALPHEUS S. On the nature and origin of the so called "spiral thread" of tracheæ. <*Amer. Nat.*, 1886, v. 20, pp. 438–442, figs.; *Journ. Roy. Micros. Soc.*, 1886, ser. 2, v. 6, pp. 789–790.

313.

1886. PACKARD, ALPHEUS S. [Review of] Korotneff: Development of the Mole cricket. <*Amer. Nat.*, 1886, v. 20, pp. 460–462, pl. 18–19.

314.

1886. PACKARD, ALPHEUS S. [Review of] Grassi: Development of the Honey-bee. <*Amer. Nat.*, 1886, v. 20, pp. 462–464.

315.

1886. PACKARD, ALPHEUS S. The origin of the spiral thread in tracheæ. A correction. <*Amer. Nat.*, 1886, v. 20, p. 558.

316.

1886. PACKARD, ALPHEUS S. Larval form of *Polydesmus canadensis*. <*Amer. Nat.* 1886, v. 20, p. 651.

317.

1886. PACKARD, ALPHEUS S. A new arrangement of the orders of Insects. <*Amer. Nat.*, 1886, v. 20, p. 808.

318.

1886. PACKARD, ALPHEUS S. The fluid ejected by notodontian caterpillars. <*Amer. Nat.*, 1886, v. 20, pp. 811–812.

319.

1886. PACKARD, ALPHEUS S. An eversible "gland" in the larva of *Orgyia*. <*Amer. Nat.*, 1886, v. 20, p. 814.

320.

1886. PACKARD, ALPHEUS S. The organ of smell in Arthropods. <*Amer. Nat.*, 1886, v. 20, pp. 889-894; 973-975. (Translated abstract from K. Kraepelin.)

321.

1886. PACKARD, ALPHEUS S. Third report on the causes of destruction of the evergreen and other forest trees in northern New England. <*Rept. Dept. Agric. for* 1885, [*part of Riley's report as Entomologist,*] 1886, pp. 319-333, figs.

1886. PACKARD, ALPHEUS S. Additions to the third report on the causes of the destruction of the evergreen and other forest trees in northern New England. <*Bull. Div. Ent. U. S., Dept. Agric.*, No. 12, 1886, pp. 17-23.

323.

1886. PACKARD, ALPHEUS S. First Lessons in Zoology. <New York: 1886, pp. 6+290, figs.

324.

1887. PACKARD, ALPHEUS S. Cave fauna of North America, with remarks on the anatomy and origin of blind forms. <*Amer. Nat.*, 1887, v. 21, pp. 82-83.

325.

1887. PACKARD, ALPHEUS S. Critical remarks on the literature of the organs of smell in Arthropods. <*Amer. Nat.*, 1887, v. 21, pp. 182-185. (Translated abstract from K. Kraepelin.)

326.

1887. PACKARD, ALPHEUS S. Hauser on the organs of smell in Insects. <*Amer. Nat.*, 1887, v. 21, pp. 279-286, pl. 13-15.

327.

1887. PACKARD, ALPHEUS S. Fourth report on Insects injurious to forest and shade trees. <*Bull. Div. Ent. U. S. Dept. Agric.*, No. 13, 1887, pp. 21-32, figs.

328.

1887. PACKARD, ALPHEUS S. Notes on certain *Psychidæ*, with descriptions of two new *Bombycidæ*. <*Ent. Amer.*, 1887, v. 3, pp. 51-52.

329.

1865. PACKARD, ALPHEUS S. Notice of an egg-parasite upon the American Tent caterpillar, *Clisiocampa americana* Harris. <*Pract. Ent.*, 1865, v. 1, pp. 14-15.

330.

1866. PACKARD, ALPHEUS S. Outlines of the study of Insects. <*Pract. Ent.*, 1866, v. 1, pp. 74-76; 94-95; 106-107, figs.

331.

1871. PACKARD, ALPHEUS S. Value of Honey-bees in fruit culture. <*West. Pomologist*, 1871, v. 2, pp. 133-134.

332.

1874. PACKARD, ALPHEUS S. The structure of Insects. <*Cult. and Count. Gentl.*, 1874, v. 39, p. 11.

333.

1874. PACKARD, ALPHEUS S. Flight, senses and growth of Insects. <*Cult. and Count. Gentl.*, 1874, v. 39, p. 22.

334.

1875. PACKARD, ALPHEUS S. The Colorado Potato-beetle and Army-worm. The Currant-worm. <*N. E. Farmer*, 1875, v. 54, No. 35, p. 1.

335.

1876. PACKARD, ALPHEUS S. The House-fly. <*Cult. and Count. Gentl.*, 1876, v. 41, p. 526.

336.

1876. PACKARD, ALPHEUS S. The Canker-worm. <*Sci. Farmer*, 1876.

337.

1878. PACKARD, ALPHEUS S. Insects injurious to the Maple. <*Sci. Farmer*, 1878.

338.

1880. PACKARD, ALPHEUS S. Insects injurious to the Cranberry. <*Trans. Wisc. Hort. Soc.*, 1880, v. 10, pp. 313–322, figs.

339.

——. PACKARD, ALPHEUS S. The House-fly, I–III. <*Youth's Companion.*

PART II.

SYSTEMATIC INDEX OF THE NEW NAMES PROPOSED.

The first number following the name refers to the number of the paper in Part I; the second, to the page where the species is first described. The following abbreviations are used in locating the types:

A. E. S. = Collection of the American Entomological Society, Philadelphia, Pa.
A. S. P. = Collection of A. S. Packard, Providence, R. I.
B. S. N. H. = Collection of the Boston Society of Natural History, Boston, Mass.
C. H. F. = Collection of C. H. Fernald, Amherst, Mass.
H. E. = Collection of Henry Edwards, New York, N. Y.
J. A. L. = Collection of J. A. Lintner, Albany, N. Y.
M. C. Z. = Collection of the Museum of Comparative Zoology, Cambridge, Mass.
N. M. = Collection of the National Museum, Washington, D. C.
P. A. S. = Collection of the Peabody Academy of Science, Salem, Mass.
S. H. S. = Collection of S. H. Scudder, Cambridge, Mass.

ARACHNIDA.

Sarcoptidæ.
Chelyletus semivorus, 64-665.*
Dermaleichus pici-pubecentis, 64-667.†
Gamasidæ.
Argas americana, 135-740.   M. C. Z.
Ixodidæ.
Ixodes albipictus, 58-366; 63-65.   M. C. Z.
  bovis, 58-370; 63-68.   M. C. Z.
  chordeilis, 63-67.   M. C. Z.
  cookei, 63-67.   M. C. Z.
  leporis palustris, 63-67.   M. C. Z.
  naponensis, 63-65.
  nigrolineatus, 63-66.   M. C. Z.
  perpunctatus, 63-68.   M. C. Z.
  unipunctatus, 58-370; 63-66.
Trombidiidæ.
Trombidium bulbipes, 132-264.
Hydrachnidæ.
Hydrachna tricolor, 80-108.
Thalassarachna, 80-107.
  verrillii, 80-107.
Oribatidæ.
Nothrus ovivorus, 64-664.*
Bdellidæ.
Bdella marina, 133-544; 278-828.   M. C. Z.

Arctiscoidea.
Macrobiotus americanus, 121-741.*
Nemastomatidæ.
Nemastoma inops, 268-203.   M. C. Z.
  troglodytes, 187-160.   M. C. Z.
Phalangidæ.
Phlegmacera, 268-203.
  cavicoleus, 268-203.   M. C. Z.
Scotolemon robustum, 187-164.   A. S. P.
Chernetidæ.
Chthonius cæcus, 268-203.   M. C. Z.
Obisium cavicola, 268-202.   M. C. Z.

MYRIAPODA.

Lysiopetalidæ.
Cryptotrichus, 261-189.
Scoterpes copei (Spirostrephon), 91-742. M. C. Z.
Spirostrephon. *See* Scoterpes.
Chordeumidæ.
Craspedosoma ocellatus (Polydesmus), 245-428.
Polydesmidæ.
Polydesmus cavicola, 187-162.   A. S. P.
  See Craspedosoma.
Pauropidæ.
Pauropus lubbockii, 74-621 ; 97-409.

* Types not preserved.   A. S. Packard.

* Based on Glover's unpublished figures.

*According to Riley this is a seasonal dimorphic form of *oxycoccana*; Fernald considers this= *minuta* Rob., and that *oxycoccana* should remain distinct.

†According to Professor Fernald this is not a Calaclysta.

‡ *Pinipestis abietivorella*, given by Grote (Check List, 1882, p. 55) as one of Packard's species, was described by Grote (Bull. U. S. Geol. Surv., v. 4, p. 701).

§ Probably varieties of *S centuriella*, Schiff.

---

* Described as a *Bombycid;* see Ent. Amer., 1885, v. 1, p. 167, for Hulst's reasons for placing here.

42

* Described from Glover's figures. *See* 328-51.
† According to Smith (Trans. Amer. Ent. Soc., v.12) none of the genera mentioned except *Glaucopis*, belong to the *Zygænidæ*. It is more convenient however, to catalogue them here.
‡ Originally described as a *Psychid*. Stretch (Ill. Zyg. & Bomby., p. 90) places near *Procris* and *Ctenucha* "chiefly because unable to assign it a more satisfactory position." Later, Packard accepted Stretch's view. Butler (Papilio, v. 1, p. 131) contends that the larva and pupa show no affinity to *Zygænidæ* or *Psychidæ*; that the structure of the imago scarcely differs from *Hurmina* of the *Dioptidæ*. Grote in his check-list (1882) follows Butler. Mr. J. B. Smith writes me that the genus is an abberent one and is "more *Lithosid* than anything else."

INDEX.

45

www.ingramcontent.com/pod-product-compliance
Lightning Source LLC
Chambersburg PA
CBHW021436090426
42739CB00009B/1501